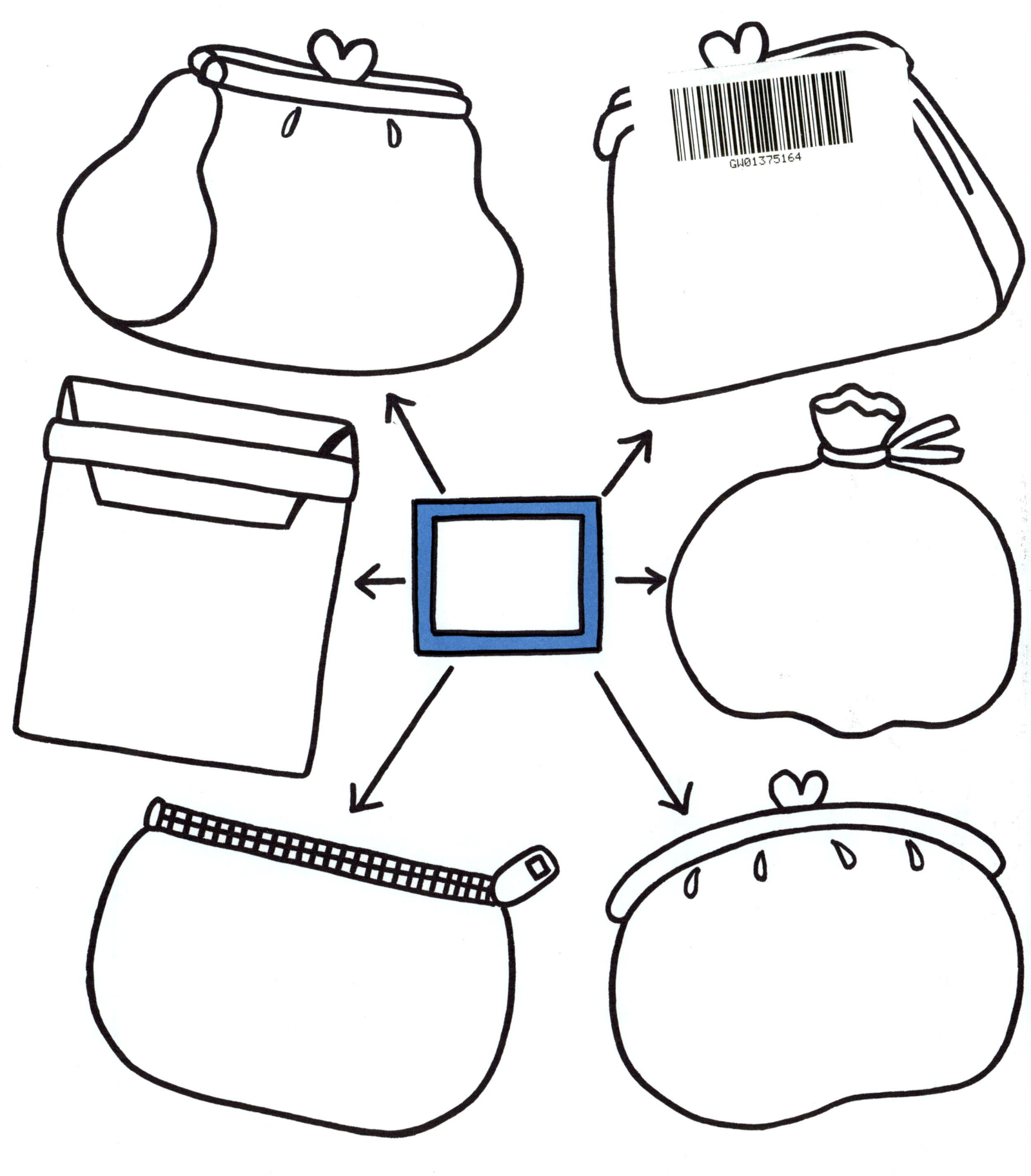

- Can work with pennies.   Notes/date:
- Can add a variety of coins.
  ....

5.2

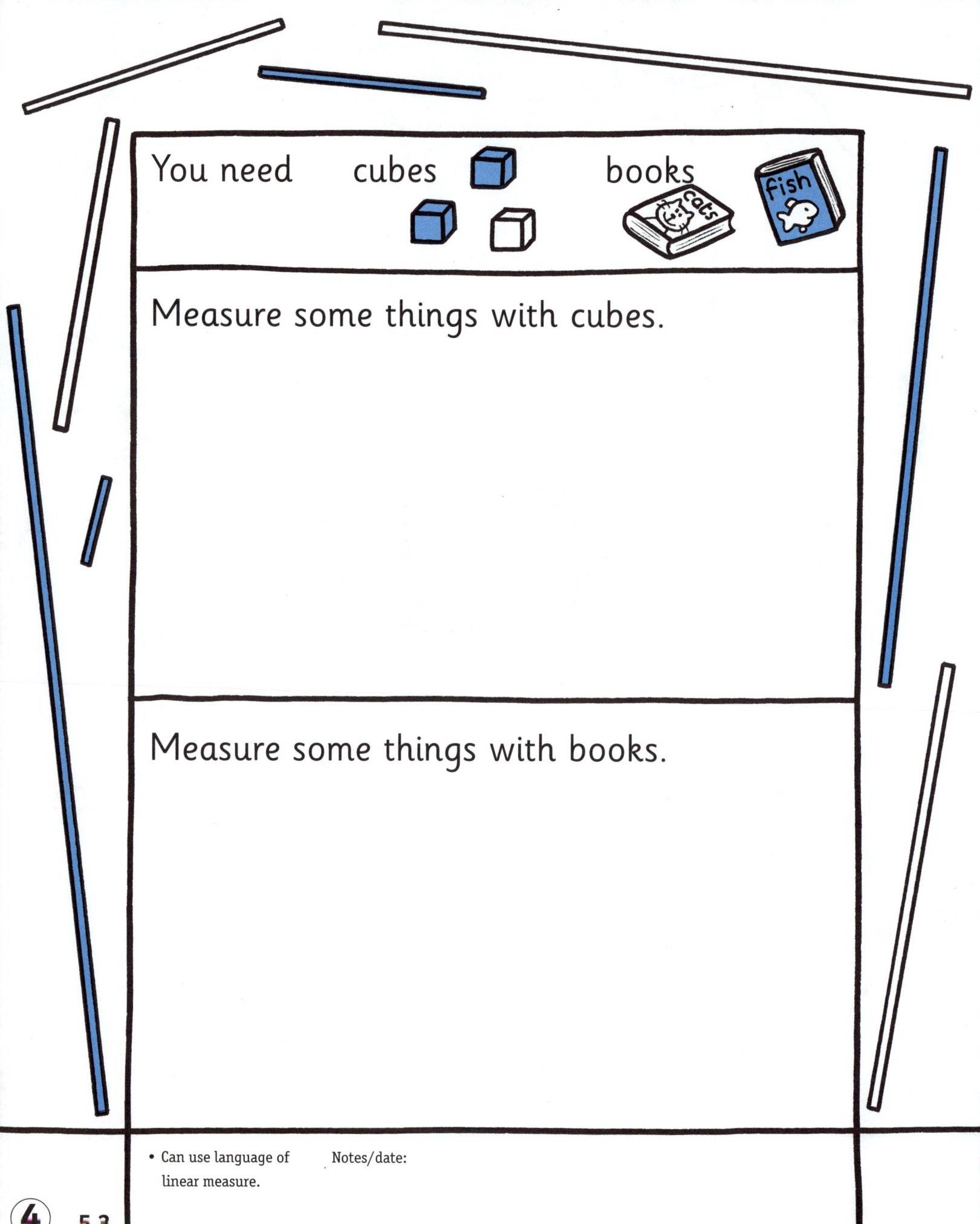>

MODULE 1 | Activity Book 5

New Cambridge MATHEMATICS

Name .... Class ....

At the baker's shop

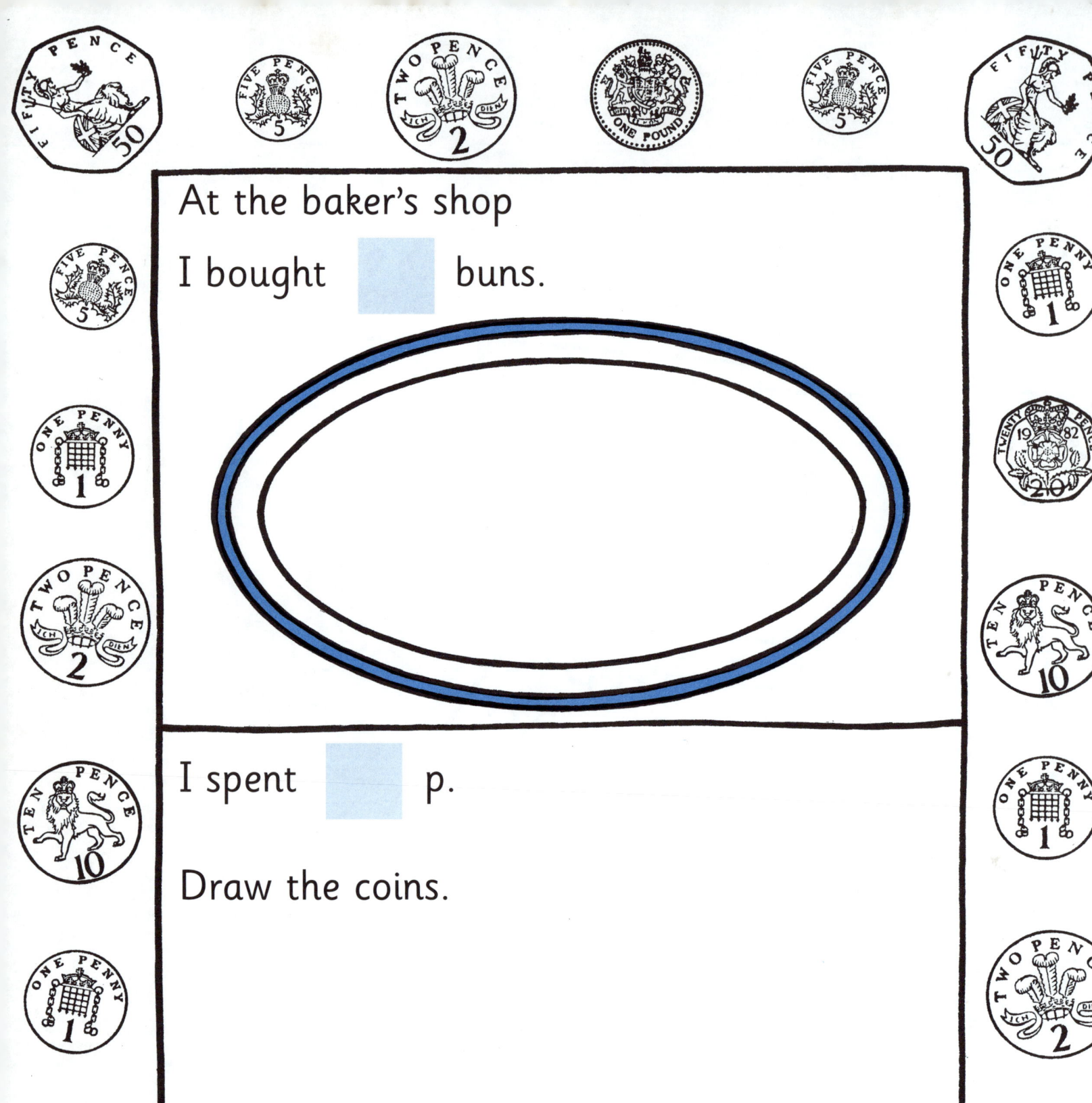

I bought ▢ buns.

I spent ▢ p.

Draw the coins.

- Can recognise coins: Notes/date:
....

Make up to 8.

☐ = 8
6 + ☐ =
2 + 1 + 3 + 2 =
4 + 4 =
I can make 8.

A day of 8

Record some of your eights.

- Can record in own way.   Notes/date:
- Enjoys exploring own ideas.

5.1 ⑤

- Can order numbers 1 to 10.   Notes/date:
- Can use ⊕ and ⊜ buttons.

5.4

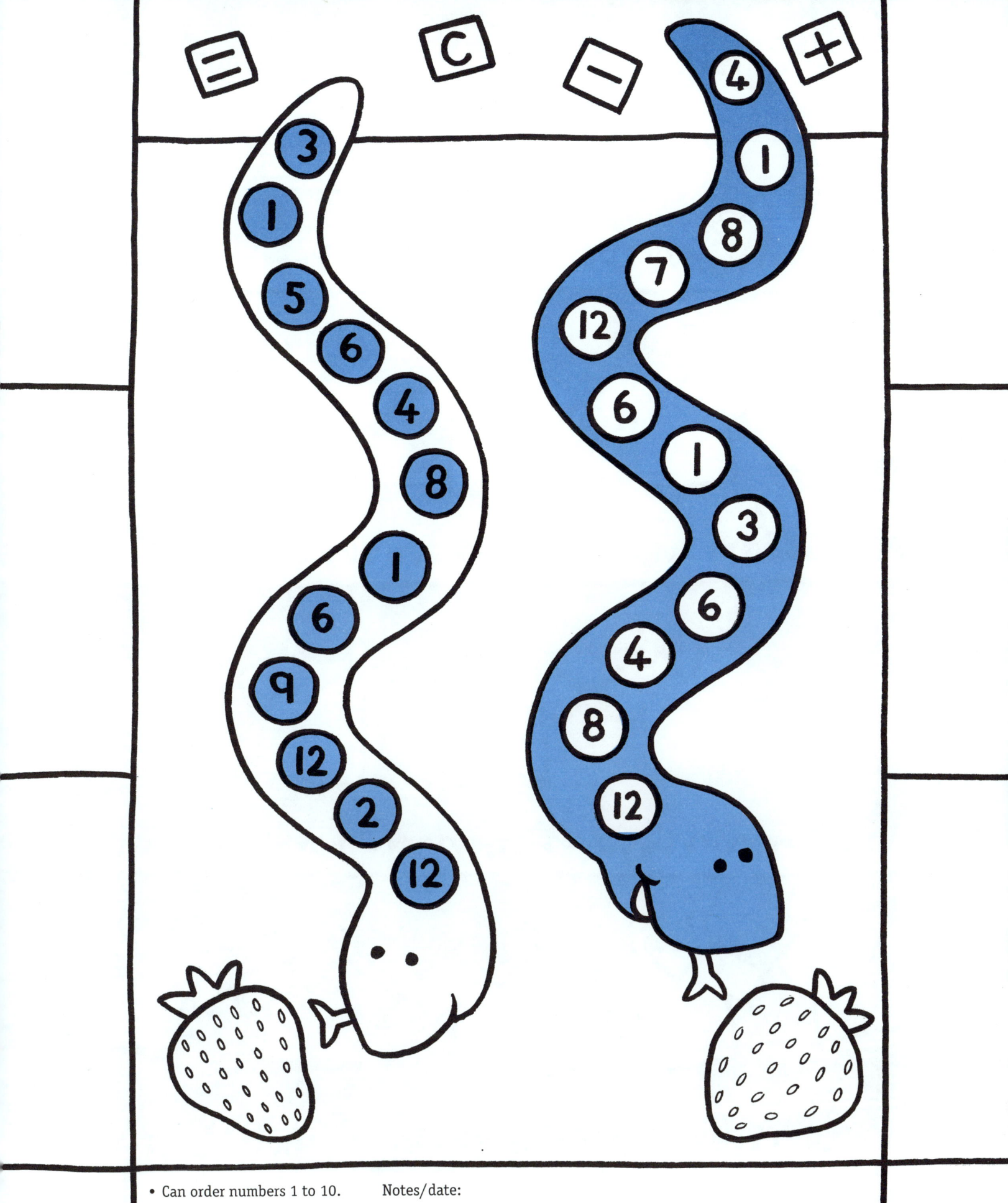

- Can order numbers 1 to 10. Notes/date:
- Can use + and = buttons.

5.4

- Can find an unknown number. Notes/date:
- Understands ideas of number patterns.

5.5

- Can find unknown numbers. Notes/date:

5.6

Draw what you made.

I measured .... with 'feet'.

I found out

Draw a picture with shapes.

- Can recognise circle, triangle, rectangle, hexagon.

Notes/date:

P.C2 11

Looking for tens

Draw round both your hands.

Find other tens in your school. Draw them.

10, 9, 8,

zero astronauts

- Confident with numbers to 10.
- Enjoys working with numbers.

Notes/date:

Finish the patterns on Emma eel and her friends.

Make patterns.

- Can design a pattern.   Notes/date:

P.C7

Published by the Press Syndicate of the University of Cambridge
The Pitt Building, Trumpington Street, Cambridge CB2 1RP
40 West 20th Street, New York, NY10011–4211, U.S.A.
10 Stamford Road, Oakleigh, Melbourne 3166, Australia

Sue Atkinson   Sharon Harrison
Lynne McClure   Donna Williams

Illustrated by Cathy Baxter

© Cambridge University Press 1995
First published 1995
Reprinted 1996
Printed in Great Britain by Scotprint Ltd,
Musselburgh

CAMBRIDGE
UNIVERSITY PRESS

ISBN 0-521-47585-6